How To Install Kodi On Firestick

––––– ✥❧✥❧✥ –––––

The Simple 2017 Updated Step By Step User Guide To Installing Kodi On Your Fire Stick

By Miles Price

Contents

Introduction

I want to thank you and congratulate you for joining me in my book, *How To Install Kodi On Firestick: The Simple 2017 Updated Step By Step User Guide To Installing Kodi On Your Fire Stick.*

This book contains proven steps and strategies on how to successfully install Kodi onto your Amazon fire stick and then find the right add-ons for the content that you want to view!

This includes step-by-step guides on how to install those add-ons and the basic configuration of Kodi to make your experience worth the effort. We will also be taking a quick look at the pitfalls of Kodi, things you need to avoid, and the legal issues that surround the use of Kodi.

Thanks again for purchasing my book, enjoy!

Chapter 1:

Introduction To Kodi

Everyone has heard the name Kodi, many of you will know it better as XBMC. If you are new to Kodi then don't worry; by the time you get to the end of this book, you will know all you need to know about Kodi, how to install it on your fire stick and what the best add-ons and builds are to use. Confused yet? Let's clear things up.

What is Kodi?

Kodi is nothing more than a piece of software. It is open source, which means that anyone can take the source code and change it, redistributing it for their own purpose. You will see this later on when we talk about different Kodi builds. Kodi is also cross-platform, which means it will work on multiple operating systems. In basic terms, Kodi is a media center that anyone can modify, use or add bits to.

Kodi was born in 2003, originally for the Xbox, hence the name XBMC – Xbox Media Center. Shortly afterward, it moved away from being Xbox-only and onto other platforms, such as Windows, Mac OS, Android, iOS, BSD, and Linux, amongst others. Kodi can turn any device it is on into a fully-fledged set-top box that can stream media, not just those on the internet but those you have stored on your network as well.'

That is nothing new these days, but Kodi stands out from all the others because it is open source. While the SBMC foundation manages the system, each of the systems in use today is very different. There is no licensing and no app stores that are strictly controlled and the means the Kodi community can develop what they want. Theoretically, that means that you can listen to or watch pretty much what you want without having to cut through all the red tape that giants like Google and Apple tie things up in.

Obviously, this leads to Kodi being very popular for the pirates but, like Bit Torrent, there is much more to Kodi than being able to freely (and illegally – more about that later) accessing material that is under a copyright. Pretty much all the infringement of copyrighted material comes via third-parties and is in no way endorsed by anyone in the Kodi community.

Does Kodi Work on All Devices?

Almost. If you go to the official download page for Kodi, you will find software downloads for most operating systems:

- Windows

- Mac OS X

- Linux

- Android

- Ubuntu

- iOS

- BSD

- Raspberry Pi 3

- Freescale IMX6

You will also find software that is dedicated for devices that will be used primarily for running Kodi and its called Kodibuntu. There are some exceptions to Kodi's compatibility, the obvious ones being Windows Phone or Windows Mobile and Chrome OS. And, to get Kodi onto your iOS device isn't quite so straightforward either.

Those of you who use windows Phone OS are out of luck because there isn't an official Kodi app for it and there are no workarounds from third-parties either. If you have a Chromebook, there are a few options and, if you are prepared to put the effort in, you can get Kodi onto it. Google is rolling out compatibility for the android Kodi app to Chromebooks so be patient and it will come to your device in good time.

For iOS, you cannot download an iOS app for Kodi. This is something to do with using third-party add-ons, which are not supported or checked by Apple. You can get it onto your iPhone or iPad with a little work and you will find the instructions to do that on the internet.

There also isn't any official app for any games consoles, which is somewhat strange considering Kodi was originally designed for the Xbox. There is a version of XBMC for the Xbox but this only works on the original console and there doesn't seem to be any support for the newer Xbox 360 or the Xbox One. There is nothing for anyone who uses a Wii, PS 3 or PS 4.

Chapter 1: Introduction To Kodi

There may be a few exceptions but you can get Kodi to run on a lot of different devices, including the Amazon fire stick. Each different device can have its own media center and its own library but there are capacities in place to make sure that all devices running Kodi can connect to shared content libraries.

What Files Does Kodi Work With?

Kodi has compatibility with just about every file type going. The only things that Kodi cannot handle are Laser Discs, Blu-Ray players that are encrypted and any analog media source.

Can It Only Be Used for Streaming Media?

Streaming is a major part of what you can do with Kodi, along with playing files from networked or local storage. To be honest, there is little that you can't get on the Kodi platform, irrespective of whether you actually subscribe to any streaming services or not. You can get the Amazon Prime add-ons if you want and, if you know Amazon, you know they don't like to have their services on any device that isn't theirs.

If you haven't let go of watching Live TV just yet, have no fear because the latest few versions of Kodi don't just allow you to stream online content or media files that have been downloaded. They also offer support for connections to Live TV broadcasters, connecting you right to the server that the content is supplied from. And, if that weren't enough, it also has DVR capabilities and a TV guide.

As with most things, there is no need to stick with the features that the stock Kodi installation gives you; you can install plenty of other add-ons to expand the capabilities and we'll be talking more about that later on.

What About Customization?

This is one of the best features of Kodi, the fact that you can customize it. I mentioned earlier that Kodi software is open-source, meaning it can be added to or customized. Most of you are not likely to go into the Kodi source code and start playing about with it but you can add some extra software so that you can get the best out of your Kodi media center.

The stock download of Kodi has no content so if you leave things as they are, you can only use it for playing media stored on networked or local storage sources. If that's what you want, then that's all good but you are missing out on so much more.

The Kodi community offers lots of different plugins and add-ons and these can be put onto your Kodi device through the Add-On Manager, something akin to an app store that you use on other systems. These plugins and add-ons give Kodi more capabilities, such as adding content from online, adding some new features and changing how Kodi works for you.

An example of this would be an add-on that provides access to catch-up and streaming services, add-ons that offer access to clients for P2P file-sharing, accessing the internet on your Kodi device, even games, albeit basic ones. We'll talk more about the best add-ons later and what they offer you in terms of content.

As for the Kodi interface, every part of it can be customized. If you don't want to get involved in that or don't feel confident enough, you can download plenty of skins to change the way your Kodi looks. Some of these come from Kodi and others are from third-parties; all of the skins are customizable so if you don't like the way it looks you can make some changes to it.

However, as much as Kodi is highly customizable, it is designed with a 10-ft interface to make it easier to use. If you don't know what that means, a 10-ft interface is a GUI (Graphical User Interface) where each of the parts of it is designed so that they work from up to 10 feet or about 3 meters away. This is designed primarily for large-screen televisions and it means that any text on the screen can be read from that distance and any input is done via remote control or a wireless keyboard and mouse.

So, Do You Need a Remote Control?

You can't operate any media center properly without a remote control and Kodi is the same. If you purchase a Kodi box or an Android TV box, generally you get the remote with it. Kodi also has an official remote app for Android and iOS meaning you have a great way of controlling Kodi from wherever you are. These apps are free and they are also open source, meaning you can customize or modify it how you want. Of course, that means there are tons of Kodi remotes to choose from but you should start with the official apps or the standard remote that comes with your box.

Now that we know what Kodi is, let's have a look at what you shouldn't do with Kodi before we move on to how to install it on your fire stick.

Chapter 2:

What Not To Do With Your Kodi Installation

Before we start to look at how to install Kodi on your fire stick, we should touch on some of the things you really shouldn't do with your installation unless you are experienced and very knowledgeable about what you are doing.

1. Messing with the settings

We humans have a very curious nature and, if there is one thing we love to do, its mess about with things, root through the settings and upgrade software whenever we can. This is fine if you know exactly what you are doing and if you are something of any expert. You also need to know how to fix what goes wrong!

Loads of people get into a lot of trouble when they start delving into the settings on their Kodi installation – there are a lot of settings and it's easy to get into trouble when you start changing things, especially when you have no idea what you are doing.

If you have never heard of hardware acceleration, debugging logs and video interlacing, even if you have heard of them but have no idea what they are and what they do, don't touch the settings. Most of the problems reported on Kodi forums come from people who have changed settings that they know nothing about, causing all sorts of problems. Typically, you will hear a Kodi user say something like, "it was all fine until I went into settings and changes some things, no I can't see any video/hear any audio" or something along those lines. There are other problems but, generally, these are all caused by the user changing a setting that stops Kodi from working as it should.

The rule of thumb is if you don't know what you are doing or you do not have experience or advanced knowledge of what you are trying to modify, stay away and leave things well alone. When you install Kodi for the first time, the settings are automatically configured for best use.

2. Upgrading Kodi to a newer version

This is another very common issue with Kodi users, especially newbies. It might seem like a great idea to upgrade when a newer version is released, after all, it has to be better, doesn't it? Sadly, that isn't always going to be the case.

Right now, the current version of Kodi is 17.1 and each of the previous versions has been pretty much the same as the last. You could have two different versions of Kodi running side by side on separate devices and you wouldn't be able to tell the difference.

Each new version contains nothing more than a few stability fixes and performance enhancements. If there are any larger changes, they are for functions that most users will never need to use and most upgrades are really only for those what stream content from another computer on the same network.

While a new Kodi version looks nice and new, if your current version works as it should do, there isn't any need to upgrade. And, doing so will likely bring about nothing but trouble for you, including video not paying, programs crashing, distortion in playback, just to mention a few.

Really, it's the same advice as for playing with the settings – don't do it if you really don't know what you are doing or if you don't need to.

3. **Deleting add-ons in Kodi**

This is another common issue – users having a problem with an add-on and jumping straight to the conclusion that the only way to solve it is to delete the add-on and then re-install it.

Three is one major problem with this brand of logic – the decision to delete is made on the thought process that all of the add-ons will always work, all of the time and will never have any issues.

Sadly, that situation only exists in a perfect world. There are times when an add-on may be causing the problems on your installation but most of the time the real cause is that the add-on is experiencing an issue that goes across the board. These issues can go from a server being down, a URL pointing problem to something as basic as not paying the server bill.

Patience is the answer when this happens. Normally, the developer of an add-on will be able to detect the problem and provide a fix for it within a day or two at the most. Too many people spend hours and hours uninstalling add-ons and reinstalling them, even uninstalling Kodi altogether and starting again. They waste time reading through reams of forums and mess about with the settings even though they don't really know what they are doing and, in the end, what it boils down to is that the add-on wasn't working for any user, not just them.

Let's look at this in a slightly different light. Let's say that you are using your computer, you open up your browser and you type in the address of a website that you visit regularly. Sometimes, you will get an error message on your screen, instead of the website and, rather than being patient, you come quickly to the conclusion that your computer is not working right or the browser has gone wrong. You delete your browser and then try reinstalling it. This won't work because it is the website that is having the problem, not your computer. You don't realize this so you take a huge step and you reformat your computer, wasting hours and hours of time that you really didn't need to because, after all that, the problem was with the website.

The lesson? Be patient; if it is a problem with the add-on, it will either be fixed or taken down.

4. **Not shutting Kodi down in the right way**

Many users have no idea of just how complex Kodi is, no matter what platform you use it on. We get so used to just hitting the home button on our mobile devices to shut down a program that we tend to think we can do the same thing with Kodi. This is natural but it isn't the right way.

Because Kodi is such a complex program, you can't just press the Home button and turn your box off. There is a lot going on in Kodi and this is why, when you press the power button and exit out of Kodi, it takes such a long time. This is annoying but the reason for it is, there is a lot of stuff going on in the background and this takes time to process; these functions are essential to the way Kodi functions and, by just shutting it off the way you do doesn't let all those processes complete.

This can lead to issues within Kodi, like interrupting updates to an add-on and breaking it. The worst that can happen, though, is that files in Kodi become corrupted and that can stop Kodi from working. If this happens, the only way out of it is to delete Kodi and start again.

The way to shut Kodi down is to click the Power button which will be at the bottom left of the screen. Then click on Exit and wait for Kodi to close down altogether and you go back to the Home screen of your device. Once you get back to the home screen, you can turn it off any way you like.

There are lots of ways that you can mess up your Kodi installation but these are the most common ones so let's recap what you need to do:

- Be patient!

- Add-ons do fail at times so have some patience and wait for the developer to provide a fix. Use another add-on while you wait, there are plenty to choose from

- Don't mess about with the settings unless you know exactly what you are doing

- Don't upgrade Kodi or delete it unless you are experienced in what you are doing

- Make sure you close Kodi down properly

Chapter 3:

How To Install Kodi On The Amazon Fire Stick

Important note – there are five methods here each of which will work with all fire sticks and Fire TV models and all builds and add-ons are working with Kodi 17.1. However, if you prefer to stick with Kodi 16.1, I will provide you with the links you need.

The first of these methods is the easiest because you won't need to use your computer to do it. All you do is install a free piece of software from the Amazon Fire TV app store and then install Kodi onto your fire stick using that. Have a scan through all five methods, though, and choose the one that suits you.

Method 1: Use Downloader to Install Kodi on Your Fire Stick

This tutorial is for Kodi 17.1 but if you prefer to stick with Kodi 16 then the link http://troypoint.com/jarvis instead of http://troypoint.com/kodistable in the Downloader program.

1. On your fire stick, open Settings and then Developer Options. Enable Apps from Unknow Sources

2. Also, enable USB or ADB Debugging as this can help with side loading that you may want to do in the future

3. Go back to your home screen and go to the Search button. In the box that comes up, type in Downloader and then click on Download or Get to install the program

4. Open Downloader and go to the box that says "Enter URL of the file you want to download" and type in the UR of the Kodi version you want to download – see URs above.

5. Click on Next

6. Now click on Download and follow the on-screen instructions for installing

7. When it has finished, open Kodi and make sure it has all installed OK.

Method 2: Use ES Explorer to Install Kodi Using Download Manager

1. Open Settings on your fire stick and open Developer Options

2. Enable Apps from Unknown Sources

3. Go back to your Home Screen and go to the Search button

4. In the search box, type in ES Explorer

5. Choose Explorer and click on Download or Get to install it

6. Open the ES Explorer app and in Download Manager, type in one of the URL's from above. Name it Kodi

7. Click on Download and Kodi will begin downloading

8. You will now see the Installation screen so click on Install and let Kodi begin installing

9. Once it has completed, go back to your home screen and go to Apps – make sure Kodi is listed

10. Open Kodi to make sure it works properly

Method 3: Use Your Computer and abdLink to Install Kodi

This works on Linux, Mac and Windows operating systems

1. Connect your fire stick to your television and follow the steps for setting it up – make sure to link your amazon account with your fire stick

2. Now go to Settings and open System, then Developer Options

3. Enable Apps from Unknown Sources and ABD Debugging

4. Next, go to About and then Network – make a note of the local IP address

5. Go to Networks and make a note of the Wireless Access Point (WAP) that your device is connected to

6. Make sure that your computer is connected to the same WAP as your fire stick

7. Now download three programs on your computer – Kodi, abdLink, and ES File Explorer

8. INTAL abdlink and connect it to your fire stick. To do this, go into Connection settings and input the IP address you noted earlier

9. Now go into abdLink and click on Install APK. Browse through for Kodi and side load it.

10. Do the same as the above for ES File Explorer

11. Open Kodi to make sure it runs OK

Method 4: Use apps2fire and Your Mobile Device to Install Kodi

This requires you to have an Android mobile device and it must be connected to the same network as your fire stick

1. Go to the settings on your fire stick, into System, and into Developer Options. Enable Apps from Unknown Sources and enable USB Debugging

2. Now go to the tab for About and hover your mouse over the button for Network – make a note of your IP address

3. On your Android mobile device, go to the Google Play Store, search for and install apps2fire

4. Also, install Kodi from the store

5. Now open apps2fire and go into Setup. Type in the IP address you wrote down earlier – this will be the IP address of your fire stick

6. Go to the menu option for Fire TV Apps and ensure that your stick is connected to your mobile device

7. Once the connection has been confirmed, click on the button that says "Local Apps"

8. Select Kodi and install it

9. Once apps2fire shows that Kodi is installed, go back to your home screen, open Kodi to see if it is working

Method 5: Use ES Explorer and AppStarter to Install Kodi

1. Open your fire stick settings and go to System and Developer Options. Enable Apps from Unknown Sources and USB Debugging

2. Follow the instructions in an earlier method to install ES File Explorer

3. Open ES File Explorer

4. Go to Favorites and click on the Add button – type in this URL – http://fireunleashed.com/as. Call it Fire or something that you will remember and click on Add

5. Under Favorites, look for the FireUnleashed source and click on it – it will open in File Explorer

6. You will see a link that says "Click Here to Install AppStarter" – click on it and then click on Open Folder

7. Now launch the .apk file to install the program

8. Now launch AppStarter and click on Updates – left side of the screen

9. Next click on Install

10. Once the installation has completed, go back to your home screen and open Kodi to make sure it works

Provided you follow the instructions carefully, you should not have any trouble with your Kodi installation

Chapter 4:

Best Kodi Add-Ons

There are plenty of add-ons to choose from for your Kodi installation and it would take forever to list them all. So, I have picked the ten best ones, the ones that will provide you with pretty much all you will need, along with full instructions on how to install each one on your fire stick for Kodi 16 and Kodi 17.1.

1. Exodus

This is by far the most popular of all the Kodi add-ons, providing access to a wide range of movies and TV shows. It rarely goes down and works very well:

Install Exodus on Kodi 16.1

- From the main Kodi screen, go to System and then File Manager

- Click on Add Source

- In the source box, click on None

- Another screen will load, type in the following URL EXACTLY as written and click on Done – http://fusion.tvaddons.ag

- The next box down asks you for a name – click on it and type in Fusion

- Before you go any further, double check the URL is correct and the name; if not, make the changes. If it is, click on OK

- Go back to the main menu and click on System and then Settings

- Now click on Add-ons

- Select Install from Zip File and a new menu will appear at the side

- Click on Fusion

- Click on Kodi-repos

- Click on English

- Click on repository.exodus-x.x.x (whichever version is available) and wait for it to install

- When you see a message on the screen that says "Add-on Enabled", click on Install from Repository

- Click on Exodus Repository

- Click on Video Add-Ons

- Click on Exodus and then install it

When it is finished, you will be able to access it from the main menu, in Videos and then Add-ons.

Install Exodus on Kodi 17.1

Note - On all add-on installations on Kodi 17.1, you must enable Unknown sources so go ahead and do this now before you attempt to install any of the add-ons listed here:

- Open Settings

- Click on System Settings

- Ensure that you are in Expert Mode and not Basic

- Click on Add-ons

- Enable Unknown Sources

- On the warning box that appears, click on Yes

- Go to the main Kodi menu

- To install Exodus on Kodi 17.1:

- Click on the System Settings icon

- Click on File Manager

- Click on Add Source (on the left)

- Click on None

- Type in this URL exactly as it is written – http://fusion.tvaddons.ag

- Click on Done and then go to the next box and click on it

- Type in Fusion as a name and click on Done

- Check that all has been typed in correctly and click on OK

- Go to the main menu and click on Add-ons

- Click on Add-on Package Installer – top left of the screen

- Click on Install from Zip File

- Click on Fusion

- On the new menu that appears, click on kodi-repos

- Click on English

- Click on repository.exodus-x.x.x (whichever version number is there) and wait while it installs.

- You will see a message on screen that says "Installed" and you may see one that says "Updated"

- Now click on Install from Repository

- Click on Exodus

- Click on Video Add-ons

- Click on Exodus and then on Install

- You will now see Exodus in the Video section on the main menu, under Add-Ons

As you can see, the instructions are pretty similar to installing on Kodi 16.1 but, for the sake of clarity, I will continue to give the instructions for both versions

2. **Phoenix**

Phoenix is another popular add-on, containing a lot of sports, TV shows and movies. It is constantly kept updated and is a great source for anyone who wants easy access to everything, including Live TV, sports, cartoons, movies, and music. Phoenix is added to by several contributors, making it an excellent source for the best streams

Install Phoenix on Kodi 16.1

- Open System from your main menu

- Click on Add-ons

- Click on Install from Zip File

- Click on Fusion

- Click on kodi-repos

- Click on English

- Click on repository.xbmchub-x.x.x (whichever version is available)

- Wait for the message that shows the add-on has been enabled and updated.

- Now click on Install from Repository

- Click on tvaddons.ag

- Click on Video add-ons

- Click on Phoenix

- Click on Install

When complete, you can access Phoenix from the main menu under Videos>Add-ons

Install Phoenix on Kodi 17.1

- Click on the System Icon

- Click on File Manager

- Click on Add Source

- Click on None

- Type in this URL exactly as it is written – http://toptutorials.co.uk/kodi/

- Click on Done and click on the next box

- Type in Top Tutorials as a name and click on Done

- Check that all is correct and click on OK

- Go to the main menu and click on Add-ons

- Click on Add-on Package Installer

- Click on Install from Zip File

- Click on Top Tutorials

- On the new menu, click on kodi-repos

- Click on English

- Click on repository.exodus-x.x.x (whichever version number is there) and wait for it to install

- You will see a message on the screen that says "TVaddons.ag Add-on Repository Enabled"

Chapter 4: Best Kodi Add-Ons

- Click on Install from Repository

- Click on TVAddons.ag

- Click on Video Add-ons

- Click on Phoenix and then Install

You will now see Phoenix in the Video section on the main menu, under Add-Ons

3. **Specto Fork**

Specto Fork is based on an old but popular repository called Genesis and anyone used to that will find the interface familiar. It has a great feature – a folder for favorites, allowing you to save your favorite programs and easily find them when you want them.

Install Specto Fork on Kodi 16.1

- Open System from the main menu

- Click on File Manager

- Click on Add Source

- Click on None

- Type in this URL exactly as written – http://kodi.filmkodi.com

- Click the next box and type in a name – Specto Fork or FilmKodi

- Click Done

- Check all is correct and then click on OK

- Go back to the main menu and click on System

- Click on Add-ons

- Click on Install from System File

- On the new menu, click on Specto Fork or FilmKodi, whatever you called it

- Click on repository.film.kodi.com

- Click on repository.filmkodi.com-x.x.x.zip (whichever version is there.)

- Wait for the message to say the add-on has been enabled and updated.

- Click on Install from Repository

- Click on Filmkodi.com repository – MRKNOW

- Click on Video add-ons

- Click on Specto Fork

- Click on Install

When complete, you can access Specto Fork from the main menu under Videos>Add-ons

Install Specto Fork on Kodi 17.1

- Click on the System Icon

- Click on File Manager

- Click on Add Source

- Click on None

- Type in this URL exactly as written – http://kodi.filmkodi.com

- Click on Done and click on the next box

- Type in FilmKodi as a name and click on Done

- Check that all is correct and click on OK

- Go to the main menu and click on Add-ons

- Click on Add-on Package Installer

- Click on Install from Zip File

- Click on filmkodi

- Click on repository.kodi.com

- Click on repository.filmkodi.com-x.x.x (whichever version number is there) and wait for it to install

- You will see a message on screen that says "MRKNOW Enabled"

- Click on Install from Repository

- Click on Filmkodi.com repository MR KNOW

- Click on Video Add-ons

- Click on Specto Fork and then Install

You will now see Specto Fork in the Video section on the main menu, under Add-Ons

4. BOB Unrestricted

BOB is another add-on that used to be with Phoenix and it contains a lot of music, films, TV shows and sports content. It allows one-click playing of streams without the need to search through lists of servers. Be prepared for some odd messages while you wait for the streams to start.

Install BOB on Kodi 16.1

- Open System from the main menu

- Open File Manager

- Click on Add Source

- Click on None and type in the following URL - http://noobsandnerds.com/portal/

- Click Done and then go to the next box

- Type in Noobs and Nerds and click on Done

- Check everything is correct and click on OK

- Back to the main menu and click on System

- Click on Settings

- Click on Add-on

- Click on Install from Zip File

- On the new menu, click on Noobs and Nerds

- Click on noobsandnerds_repo.zip

- Wait for the Enabled or Updated message on screen and then click on Install from Repository

- Click on noobsandnerds repository

- Click on Video add-ons

- Click on BOB and then Install

When the installation is complete, you can access BOB form Videos>Add-ons on the main menu

Install BOB Unrestricted on Kodi 17.1

- Click on the System Icon

- Click on File Manager

- Click on Add Source

- Click on None

- Type in the following URL - http://noobsandnerds.com/portal/

- Click on the next box and type in Noobs and Nerds as a name

- Click Done, check that all is OK and click on OK

- Now click on Add-ons and click the Package Installer

- Click Install from Zip File

- Click on Noobs and Nerds from the menu on the left

- Now click on noobsandnerds_repo.zip

- Wait for the "NoobsandNerds Repository Enabled" message

- Next click on Install from Repository

- Click on NoobsandNerds Repository

- Click on Video Add-ons

- Click on BOB

- Click on Install

When finished, you can access it from Video>Add-ons on the main menu

5. **Zen**

Zen is like Exodus, with loads of movies and TV shows but, it has more settings and a way of setting up a favorites folder.

Install Zen on Kodi 16.1

- On the main menu, click on System

- Click on File Manager

- Click on Add Source

- Click on None

- Type in this URL - http://noobsandnerds.com/portal

- Click on OK and then click the next box down

- Type in Noobs and Nerds

- Click on Done and then check everything is correct

- Click on OK

- Now go back to the main menu and click on System

- Click on Settings

- Click on Add-ons

- Click on Install from Zip File

- From the menu on the left click on noobsandnerds_repo.zip

- Wait for the message that says "NoobsandNerds Repository enabled"

- Then click on Install from Repository

- Click on NoobsandNerds Repository

- Click on Video Add-ons

- Click on Zen

- Click on Install

You will be able to access it from the main menu under Video>Add-ons

Install Zen on Kodi 17.1

- Click on the System Icon

- Click on File Manager

- Click on Add Source

- Click on None

- Type in this URL - http://noobsandnerds.com/portal

- **Click on Done and then go to the next box**

- **Type in Noobs and Nerds as a name and click on Done**

- **Check everything is correct and then click on OK**

- **Back to the main menu and click on Add-ons**

- **Click on Package Installer**

- Click on Install from Zip File

- Click on Noobs and Nerds from the left menu

- Click on noobsnerds_repo.zip

- Wait for the Noobs and Nerds Enabled message to appear

- Now click on Install from Repository

- Click on NoobsandNerds Repository

- Click on Video Add-ons

- Click on Zen and then click on Install

Zen will now be accessible from the Video>Add-ons section of the main menu

6. UK Turks Playlist

UK Turks is a nice playlist with a bit of everything, including TV shows, live TV, movies. Sports, documentaries, and cartoons. It features on-click playing and no need to search through server lists for decent feeds.

Install UK Turks Playlist on Kodi 16.1

- Go to System

- Click on File Manager

- Click on Add Source

- Click on None

- Type in this URL - http://kodi.metalkettle.co/

- Click on Done and then go to the next box

- Type in Metal Kettles and click on Done

- Check everything is correct and click on OK

- From the main menu click on System

- Click on Settings

- Click on Add-ons

- Click on Install from Zip File

- Click on Metal Kettles

- Click on repository.metalkettles-x.x (whichever version is available)

- Wait for the Add-on Enabled message to appear

- Now click on Install from Repository

- Click on MetalKettles Add-on Repository

- Click on Video Add-ons

- Click on UK Turk Playlist and it will download

When the installation is complete, you can access it from Videos>Add-Ons on the main menu

Install UK Turks Playlist on Kodi 17.1

- Click on the System Icon

- Click on File Manager

- Click on Add Source

- Click on None

- Type in this URL – http://kodi.metalkettle.co

- Click on Done and then go to the next box

- Type in Metal Kettles as the name and click on Done

- Check everything is correct and click on OK

- From the main menu, click on System

- Click on Settings

- Click on Add-ons

- Click on Install from Zip File

- Click on Metal Kettles

- Click on repository.metalkettle-x.x (whichever version is available)

- Wait for the message that says it is enabled and/or updated

- Now click on Install from Repository

- Click on MetalKettles Add-on Repository

- Click on Video Add-ons

- Choose UK Turk Playlist and it will install

You can now access it from the Video section on the main menu

Installing UK Turks Playlist Manually

If you struggle with the above method or it doesn't work properly, you can install it manually:

- Go to http://kodi.metalkettle.co

- Download repository.metalkettle_x.x.x.zip – save it to C:/ as Kodi cannot recognize a folder or a directory

- In Kodi, click on System

- Click on Settings

- Click on Add-ons

- Click on Install from Zip File

- Locate the file you downloaded and click on it

From here you can follow the instructions for the methods above

7. **SportsDevil**

Aa you would expect from the name, SportsDevil is the home for sports but it does have other content on it, including live streams from Fox News, Cartoon Network, and many other channels. The live streams won't always work very well as it will depend on the number of users attempting to use them at the same time.

Install SportsDevil on Kodi 16.1

If you already have Fusion installed on your system you can ignore the first part of this tutorial:

- Open System from the main menu

- Click on File Manager

- Click on Add Source

- Click on None

- Type in this URL: http://fusion.tvaddons.ag

- Click on Done and then click on the next box down

- Type in Fusion and click Done

- Check that everything is correct and click on OK

- Go to the main menu and click on System

- Click on Settings

- Click on Add-ons

- Click on Install from Zip File

- Click on Fusion on the left menu

- Click on kodi-repos

- Click on English

- Click on repository.unofficialsportsdevil-x.x.x (depending on the version available).zip

- Wait for the add-on enabled message

- Click n Install from Repository

- Click on Unofficial SportsDevil Repository

- Click on Video Add-ons

- Click on SportsDevil

- Install it

SportsDevil can now be accessed from the main menu under Video>Add-ons

Install SportsDevil on Kodi 17

- Click on the System icon

- Click on File Manager

- Click on Add Source

- Click on None

- Type in this URL: http://fusion.tvaddons.ag

- Click on Done and then on the next box down

- Type in Fusion and click on Done

- Check everything is correct and then click on OK

- Now go back to the main menu and click on Add-Ons

- Click on Package Installer

- Click on Install from Zip File

- Click on Fusion

- Click on kodi-repos

- Click on English

- Click on repository.unofficialsportdevil-x.x.x (Depending on the version available).zip

- Wait for the Sports Unofficial Sports Devil Repo Enabled message

- Now click on Install from Repository

- Click on Unofficial Sports Devil Repository

- Click on Video Add-ons

- Click on Sports Devil

- Click on Install

SportsDevil will now be available from your main menu, under Video>Add-ons

8. SALTS

SALTS is an acronym for Stream All The Sources and it is a somewhat interesting one, as it was developed to try and fix the issue of sources that are unreliable. You can use paid services if you choose; these are called Real Debrid Accounts but there is also the free option which works just as well

Install SALTS on Kodi 16.1

If you already have Fusion installed, ignore the first part of the instructions:

- From your main menu, click on System

- Click on File Manager

- Click on Add Source

- Click on None

- Type in this URL - http://fusion.tvaddons.ag

- Click on Done and then click the next box

- Type in Fusion and click on Done

- Check all is correct and then click on OK

- Back to your main menu and click on System

- Click on Settings

- Click on Add-ons

- Click on Install from Zip File

- Click on Fusion on the left menu

- Click on kodi-repos

- Click on English

- Click on repository.tknorris.release-x.x.x.zip

- Wait for the message that says the add-on is enabled

- Next, click on Install from Repository

- Click on tknorris Release Repository

- Click on Video Add-ons

- Click on Stream All The Sources

- Click on Install

Once the installation is complete, you can access SALTS from the main menu, under System and Add-ons

Install SALTS on Kodi 17

- Click on the System icon

- Click on File Manager

- Click on Add Source

- Click on None

- Type in this URL - http://fusion.tvaddons.ag/

- Click on Done and then on the next box

- Type in Fusion and click on Done

- Check it is all correct and click on OK

- From the main menu click on Add-ons

- Click on Package Installer

- Click on Install from Zip File

- Click on Fusion from the menu on the left

- Click on kodi-repos

- Click on English

- Click on repository.tknorris.release-x.x.x.zip

- Wait for the add-on to be enabled

- Now click on Install from Repository

- Click on tknorris release repository

- Click on Video Add-ons

- Click on SALTS or Stream All The Sources

- Click on Install

When it has installed you can access it from System and Add-ons from the main menu

9. 1Channel

1Channel has been in existence for some time now and is still a favorite. It has pretty much the same content as Exodus so if one goes down you can watch the other. It has a basic interface that doesn't show so much information about shows or movies.

Install 1Channel on Kodi 16.1

If you already have Fusion installed, ignore the first part of the instructions

- Open System from your main menu

- Click on File Manager

- Click on Add Source

- Click on None

- Type in this URL - http://fusion.tvaddons.ag/

- Click on Done and then click the next box

- Type Fusion and click Done

- Click on OK when you have checked that all is correct

- Go back to the main menu and click on System

- Click Settings

- Click Add-ons

- Click Install from Zip File

- Click Fusion on the menu on the left

- Click kodi-repos

- Click English

- Click repository.tknorris.release-x.x.x.zip

- Wait for the add-on installed message

- Now click on Install from Repository

- Click tknorris repository

- Click on Video Add-ons

- Click on 1Channel

- Click on Install

After installation, you can access it from the Video>Add-ons section on the main menu

Install 1Channel on Kodi 17

- Click on the System icon

- Click on File Manager

- Click on Add Source

- Click on None

- Type in this URL - http://toptutorials.co.uk/kodi/

- Click on Done and then on the next box

- Type in Top Tutorials and click on Done

- Check it is all correct and click on OK

- From the main menu click on Add-ons

- Click on Package Installer

- Click on Install from Zip File

- Click on Top Tutorials from the menu on the left

- Click on repos

- Click on repository.tknorris.release-x.x.x.zip

- Wait for the add-on to be enabled

- Now click on Install from Repository

- Click on tknorris release repository

- Click on Video Add-ons

- Click on 1Channel

- Click on Install

When it has installed you can access it from Video and Add-ons from the main menu

How to Install 1Channel Manually

- Download tknorris repo and remember where you save it to – C:/ is best

- Open Kodi and go to system

- Click on Settings

- Click on Add-ons

- Click on Install from Zip File

- Find the file you downloaded and click on it

- Wait for the installation to be completed and then click on Install from Repository

- Click on tknorris repository

- Click on Video Add-ons

- Click on 1Channel and wait for it to be installed

As above, you can access it via Video and Add-ons

10. Silent Hunter

Silent Hunter is a new one with lots of content that includes 4K and new releases, as well as genre and stand up. You don't need to go through server lists because movies play with one click

Install Silent Hunter on Kodi 16.1

- From your main menu, click on System

- Click on File Manager

- Click on Add Source

- Click on None

- Type in this URL - http://toptutorials.co.uk/kodi/

- Click on Done and then click the next box

- Type in Top Tutorials and click on Done

- Check all is correct and then click on OK

- Back to your main menu and click on System

- Click on Settings

- Click on Add-ons

- Click on Install from Zip File

- Click on Top Tutorials on the left menu

- Click on repos

- Click on repository.silenthunter.zip

- Wait for the message that says the Silent Hunter add-on is enabled

- Next, click on Install from Repository

- Click on Silent Hunter Repository

- Click on Video Add-ons

- Click on Silent Hunter

- Click on Install

Once the installation is complete, you can access Silent Hunter from the main menu, under Videos and Add-ons

Install Silent Hunter on Kodi 17

- Click on the System icon

- Click on File Manager

- Click on Add Source

- Click on None

- Type in this URL - http://toptutorials.co.uk/kodi/

- Click on Done and then on the next box

- Type in Top Tutorials and click on Done

- Check it is all correct and click on OK

- From the main menu click on Add-ons

- Click on Package Installer

- Click on Install from Zip File

- Click on Top Tutorials from the menu on the left

- Click on repos

- Click on repository.silenthunter.zip

- Wait for the Silent Hunter add-on to be enabled

- Now click on Install from Repository

- Click on Silent Hunter

- Click on Video Add-ons

- Click on Silent Hunter

- Click on Install

When it has installed you can access it from Video and Add-ons from the main menu

Keep in mind that, like a lot of software and programs, add-ons may not be compatible with all hardware and all versions of Kodi. Most add-ons have been updated to work on most versions and hardware but, sometimes, there will be a problem. If an add-on cannot be installed, check which version of Kodi you are running and what hardware you have. Information from the internet will tell you if the add-on is compatible.

Chapter 5:

Best Kodi Builds

As with add-ons, there are also plenty of different Kodi builds that you can install. A Kodi build is where a developer has taken the source code and reinvented it to produce their own version of Kodi. Here we are going to look at some of the very best builds, along with full instructions on how to install them. But first, before you install a build, you should do a fresh start to give you a clean platform to start from. A fresh start will take Kodi back to factory settings; here's how to do it:

You need the Fusion repository – instructions on how to install that can be found in the previous chapter. Once installed:

- Click on System and then Settings

- Click on Add-ons

- Click on Install from Zip File

- Click on Fusion

- Click on begin-here

- Click on plugin.video.freshstart-x.x.x.zip

- Wait for the Add-on Enabled message and then go to the main menu

- Click on Programs

- Click on Fresh Start

- A message will appear asking if you want to restore Kodi to default; click on Yes

- Your Kodi installation will need to reboot and you will be back to a fresh install

How to Install a Kodi Build

I will be giving instructions with each of the builds but they tend to follow the same path, as follows:

- You add the source into file manager

- You install the plug-in for the repository but only in some cases

- You install the program add-on

- You install the build

- You restart Kodi. You will be asked to force-close Kodi and restart it – usually, you do this by simply unplugging the device and then reconnecting it

Best Kodi Builds for 2016/2017:

Boom Shakalaka

This build comes with lots of add-ons already installed and runs very smoothly. Packages are cleaned whenever the build is restarted and the cache is also cleared to keep it running properly. To update Boom Shakalaka, simply go into System from the Home screen and click on Update Build.

- From Kodi home screen, click on System

- Click on File Manager

- Click on Add Source

- Click on None and type in this URL – http://dimitrology.com/repo

- Click Done and then give the source a name

- Now go to the main menu and click on Systems and then Settings

- Click on Add-Ons and then Install from Zip File

- Click on the source and then on plugin.video.wiardology.zip

- Now go to Programs and Click on Dimitrology Builds

- Install Boom Shakalaka

- Restart Kodi and the new build is installed

Beast Encore Build

Beast provides access to loads of content which is all organized properly and very easy to access. There is something for everyone in this build, from movies and TV shows to sports, Live TV, and cartoons. Add-ons include SALTS and Exodus amongst many others.

Before you can install the Beast build, on your computer go to https://thebeast1.com/signup2/register.php?uksite   to register. Scroll down the page to find the proper registration section. Once you have registered, click on Log In and then look for the box that has the name you used to register with. Click Verify Here and you will see Yes to confirm. Now you can to Kodi on your fire stick.

- Open System and then Settings

- Open File Manager and Add Source

- Click on None

- If you are from the UK, type in http://thebeast1.com/repo; if you are from the US, type in http://thebeast2.com/repo. If you are not from the UK or US, choose the one that is nearest to you

- Name the source and go back to the main menu

- Now go to System>Settings and click on Add-ons

- Click on Install from Zip File and then click on the source

- Click on plugin.video.beast.zip

- From Programs, click on The Beast Wizard

- If you have a remote, click on the Menu button; if not, press the 'c' button on your keyboard

- Click on Add-on Settings and type in the credentials you registered on the site with earlier. Click on OK

- Now open the Beast Wizard and click on The Beast Encore build

- Restart Kodi to get the new build

Tomb Raider Build

Tomb Raider comes with a lot of add-ons that let you get straight to your favorite movies and TV shows, as well as live streams and documentaries. There are some other add-ons that make this stand out, though. Under Extras Zone you will find the UK Vehicle Check, letting you check out any IK vehicle status and, for those of you who love fitness, check out the Gym World add-on.

- From the main menu, click on System>File Manager

- Click on Add Source and type in this URL - http://tombraiderbuilds.netne.net/repo/

- Give the source a name and go back to your main menu.

- Click on System>Settings>Add-ons and then on Install from Zip File

- Click on the source and then on repository.tombraider.official-x.x.zip

- Now click on Install from Repository

- Click on Tomb Raider Repo

- Click on Program Add-ons

- Click on Tomb Raider Builds Wizard and then Install

- Wait for the Add-on Enabled message

- Go to Programs and click on Tomb Raider Build Wizards to open it

- Click on Tomb Raider Builds Here and then on Nemesis

- Install it and then click on Tomb Raider 2.48 Jarvis and wait while it installs

Ares Build

Ares build contains all your favorite content, although it is a small build. This makes it quicker but still gives you all your favorite movies and TV shows, sport, and Live TV as well as a Kids Zone.

- From the main menu, go to System>File Manager

- Click on Add Source and then on None

- Type in this URL -http://www.areswizard.co.uk/ and then name it

- Go back to the main menu and click on System>Settings

- Click on Add-ons and then Install from Zip File

- Click on the source and then on script.areswizardx.x.xx.zip

- Go back to Programs and click on Ares Wizard

- Click on Browse Builds and then click on The Ares Build

- Now you need a PIN so follow the instructions on the screen to get it and then click on Enter Pin

- Input the pin and then click on The Ares Build

- Click on Install and wait for the installation to complete

- Reboot Kodi for the build to be implemented

Nemesis Build

Nemesis Build is organized very well with an interface that doesn't allow you to become distracted when you are looking for something. It is a simple build, small and quick with lots of TV shows, movies, sports and kids' stuff.

- From your main menu go to System>File Manager

- Click on Add Source and then on None

- Type in this URL – http://kodiuk.tv/repo and then name it

- Back to the main menu, click on System and then Settings

- Click on Add-ons and then Install from Repository

- Click on KodiukTV Repo

- Click on Program and then Add-ons

- Click on KodiUK TV Wizard

- From Programs, open the Wizard and click on Kodi UK TV Builds

- Click on Nemesis

- Reboot your installation for Nemesis to be installed

Void Build

Void is full of fabulous add-ons but it truly stands out with the section called Her Place. This is a part of the Phoenix add-on that is aimed at women. There is also a Gaming category with loads of gaming videos from sources such as YouTube and Twitch

- Open System from your main menu

- Click on File Manager and Add Source

- Click on None and type in this URL - http://wookiespmc.com/wiz/

- Give the source a name and then go back to the main menu

- Click on System>Settings and Install from Zip File

- Click on the source and then click on Click me – succumb to The Wookie.zip and then on Install

- Back to Programs, click on Programs and then Wookie

- Click on Community Builds and The Void Build

- Click on Install and when the process has completed, reboot Kodi

Razer Blue Family Build

Razor Blue Family provides a wide range of content for all the family. You can choose from documentaries, sports, TV shows, movies and kids content. It isn't a big build but it is quick and also includes both 4K and 3D content for those that can watch it.

- From the Kodi main menu, click on System and then File Manager

- Click on Add Source and None

- Type in this URL - http://areswizard.co.uk and then give the source a name

- Go back to the main menu and click on System and then Settings

- Click on Install from Zip File and then click on Ares (or whatever name you gave it)

- Click on script.areswizardx.x.xx.zip to install it

- On the Main menu click on Programs

- Click on Ares Wizard and then Browse Builds

- Click on Razer Builds

- Click on Razer Blue Family

- Click on Install

- Wait for the process to complete and then reboot Kodi for the build to be implemented

Turk T2k Without Arcade

Turk 2K is a small build but it is packed out with content from movies to music, from sports to live TV. There are also lots of animated shows under a separate category.

- From your main Kodi menu, open System

- Open File Manager

- Click on Add Source and then on None

- Type in this URL: http://thecommunityrepo.netai.net/zip/ and then give the source a name

- Go back to the main menu and click on System

- Click on Settings and then on Add-ons

- Click on Install from Zip File and then click the source

- Click on repository.communityrepo-x.x.x.zip

- Staying in Settings, click on Install from Repository

- Click on The Community Repository

- Click on Program Add-ons

- Click on The Community Wizard

- Click on Install and wait until you see the add-on enabled message

- From the main menu, click on Programs

- Click on The Community Wizard and then on Community Wizard Builds

- Click on Turk T2K Without Arcade Build

- Lastly, click on Community Wizard Standard Install

- When the installation has completed, reboot Kodi

Gears of War Build

Gears of War is a neat build that offers basic movie, sports and TV show packages with some kids' stuff added in as well. It's ideal for those who don't want anything complicated and just want basic access.

- From your main Kodi menu, click on System, and then on File Manager

- Click on Add Source and then click on None

- Type in this URL - http://targetbuilds.net/repo

- Give the source a name

- Go back to the main menu and into System

- Click on Settings and then on Add-ons

- Click on Install from Zip File

- Choose Gears of War or whatever name you gave it and then click on plugin.video.targetin1080pwizard.zip

- Back to your main menu and into Programs

- Click on Target1080p Wizard and then on Gears of War Build

- Wait for the installation to complete and then reboot Kodi for it to be implemented.

Chapter 5: Best Kodi Builds

Apollo AiOne Build

Apollo AiOne is a comprehensive and deep build with loads of enjoyable content. What really stands out is a section dedicated to both 4K and 3D content, ideal for those that can watch it. There is also a health category.

- From the main Kodi menu, go to System and then File Manager

- Click on Add Source

- Click on None and then type in this URL - http://apkb.netne.net/repo/

- Give the source a name

- Go back to the main menu and click on System and then Settings

- Click on Add-ons and Install from Zip File

- Click on the source and then on plugin.program.apollo2.zip

- Back to the main menu and click on Programs

- Click on Apollo Wizard and then on Apollo AiOne

- Wait for the build to be installed and then reboot Kodi

Simplify Build

Simplify is a small build with a pretty good amount of content. You get access to loads of top movies, kids shows, TV shows and sports. You also gain access to 4Horsemen which offers up both 4K and 3D content.

62

- From the main Kodi menu, go to System

- Click on File Manager

- Click on Add Source and then on None

- Type in the following URL - http://lentechtv.com/lentechtv/ and then give the source a name

- Now go back to the main menu and click on System

- Click on Settings

- Click on Add-ons and then Install from Zip File

- Click on the source and then on plugin.video.LentechTVx.x.zip

- Back to the main menu, click on Programs and then on LentechTV Build Installer

- Click on Simplify and wait for the build to install

- Reboot Kodi

Evolution Build

Evolution is a neat build with loads of add-ons included. You can have access to sports, Live TV, movies, shows and much more besides. There is one thing that makes Evolution stand out from other builds and that is that it splits high-quality 3D and 1080p HD films away and puts them in their own section, making them easily accessible.

- From the Kodi main menu, click on System

- Click on File Manager

- Click on Add Source and then click on None

- Type in this URL – http://repo.hackmykodi.com/wizards/

- Give the source a name

- Go back to the main menu and click on System

- Click on Settings and then on Add-ons

- Click on Install from Zip File and then click on the source

- Click on plugin.video.evolutionwizard.zip

- From the main menu, click on Programs

- Click on evolutionwizard

- Click on Evolution

- Wait for the build to install and then reboot Kodi

Spinz TV Premium Lite Build

The name says it all – Spinz is a lightweight build that doesn't use too many resources. That said, there is still loads of content, from live TV to TV shows, to sports, movies and kids content.

- From the Kodi main menu, click on System

- Click on File Manager

- Click on Add Source and then click on None

- Type in this URL – http://spinztvrepo.com

- Give the source a name

- Go back to the main menu and click on System

- Click on Settings and then on Add-ons

- Click on Install from Zip File and then click on the source

- Click on plugin.program.SpinzTV-x.x.x.zip

- From the main menu, click on Programs

- Click on SpinzTV

- Click on SpinzTV Builds and then Spinz-TV Premium Lite Build

- Wait for the build to install and then reboot Kodi

Aqua Build

Aqua is another light build but one that is packed with content. It will work very well on those devices that do not stick.have much processing power, such as the fire

- From the Kodi main menu, click on System

- Click on File Manager

- Click on Add Source and then click on None

- Type in this URL –
  http://thecommunityrepo.netai.net/zip/

- Give the source a name

- Go back to the main menu and click on System

- Click on Settings and then on Add-ons

- Click on Install from Zip File and then click on the source

- Click on repository.communityrepo-x.x.x.zip

- Click on Install from Repository and click on The Community Repository

- Wait for the add-on to be enabled

- From the main menu, click on Programs

- Click on The Community Wizard and then on Community Wizard Builds

- Click on Aqua and then on Standard Install

- Wait for the build to install and then reboot Kodi

Saints Sky One + 1 Build

Saints build has something to suit everyone from the latest movie releases, to your favorite sports or TV shows. There is a section called Man Cave especially for the men and Her Place for women

- From the Kodi main menu, click on System

- Click on File Manager

- Click on Add Source and then click on None

- Type in this URL – http://saintskodi.uk/

- Give the source a name

- Go back to the main menu and click on System

- Click on Settings and then on Add-ons

- Click on Install from Zip File and then click on the source

- Click on Click-here-for-Saints.zip

- From the main menu, click on Programs

- Click on Saints Wizard and then on SAINTS SKY ONE + 1

- Wait for the build to install and then reboot Kodi

Wullies Mini Build

Wullies is another of the light builds making it ideal for the fire stick and other devices that don't have a great deal of processing power. It uses little in the way of resources and, whenever it is started, it will delete files that are not needed. There is a category for 4Horsemen, which contains 3D and 4K content.

- From the Kodi main menu, click on System

- Click on File Manager

- Click on Add Source and then click on None

- Type in this URL – http://wookiespmc.com/wiz/

- Give the source a name

- Go back to the main menu and click on System

- Click on Settings and then on Add-ons

- Click on Install from Zip File and then click on the source

- Click on Click me – succumb to The Wookie.zip

- From the main menu, click on Programs

- Click on Wookie and then on Community Builds and then Wullies Mini Build

- Wait for the build to install and then reboot Kodi

Horus Build

Horus is yet another build that is light on resources and idea for those devices with less powerful processing power. That said, there is still loads of content to be watched through the many add-ons that are included.

- From the Kodi main menu, click on System

- Click on File Manager

- Click on Add Source and then click on None

- Type in this URL – http://echocoder.com/repo

- Give the source a name

- Go back to the main menu and click on System

- Click on Settings and then on Add-ons

- Click on Install from Zip File and then click the source

- Click on repository.echo-x.xx.zip

- Click on Install from Repository and click on Echo Repository and then Program Add-ons

- Click on Echo Wizard and then Install

- Wait for the add-on to be enabled

- From the main menu, click on Programs

- Click on Echo Wizard and then on Official Echo Builds, then click on Horus

- Click on Install

- Wait for the build to install and then reboot Kodi

Fire TV Build

As well as being a resource-light build, you also get access to loads of content from music to movies, from sports to live TV and kids content. There is also the Her Place category aimed at women

- From the Kodi main menu, click on System

- Click on File Manager

- Click on Add Source and then click on None

- Type in this URL – http://firetvguru.net/fire

- Give the source a name

- Go back to the main menu and click on System

- Click on Settings and then on Add-ons

- Click on Install from Zip File and then click on the source

- Click on repository.firetvguru.zip

- Click on Install from Repository and click on Fire TV Guru Repo

- Click on Program Add-ons

- Click on Fire TV Wizard and then Install

- Wait for the add-on to be enabled

- From the main menu, click on Programs

- Click on Fire TV Wizard and then on Builds, then click on Fire TV Build

- Click on Standard Install

- Wait for the build to install and then reboot Kodi

Pulse CCM Build

Pulse is an incredibly popular build with loads of built-in add-ons to give you the best experience. There is also a dedicated section for Workout, ideal for those who are into fitness.

- From the Kodi main menu, click on System

- Click on File Manager

- Click on Add Source and then click on None

- Type in this URL – http://www.areswizard.co.uk/

- Give the source a name

- Go back to the main menu and click on System

- Click on Settings and then on Add-ons

- Click on Install from Zip File and then click on script.areswizardx.x.xx.zip

- From the main menu, click on Programs

- Click on Ares Wizard and then on Browse Builds, then click on Pulse

- Now you need a PIN so follow the instructions on the screen to get the pin

- Click on Enter Pin and input the number

- Click on Pulse and then Install

- Wait for the build to install and then reboot Kodi

No Limits Magic Build

This is one of the most complete and powerful builds, updated regularly and including loads on the best add-ons, including Live TV IPTV. There is something for everyone here, including sports, movies, music and TV shows.

- From the Kodi main menu, click on System

- Click on File Manager

- Click on Add Source and then click on None

- Type in this URL – http://kodinolimits.com/kodi/. **If this doesn't work you can try http://kodinolimits.srve.io/kodi/**

- Give the source a name

- Go back to the main menu and click on System

- Click on Settings and then on Add-ons

- Click on Install from Zip File and then click on the source

- Click on **plugin.video.nolimitswizard.zip**

- From the main menu, click on Programs

- Click on No Limits Wizard and then on Browse Builds, then click on one of the builds

- Wait for the build to install and then reboot Kodi

Chapter 6:

Legal Issues

There has been a lot of talk in recent times over whether Kodi is legal to use or not and it all comes down to the way you use it. If you are using it to stream TV, you might be facing a whole world of security and legal pain.

Although Kodi is one of the best platforms for streaming, it does raise a number of legal questions and it does have some security vulnerabilities that you do need to protect yourself from. Mostly, you can get around these by installing a VPN.

When Kodi first came out, under the name of XBMC, it was designed for gamers to play video games and watch TV shows as well as streaming media. Now it has expanded to be used on lots of different operating systems and hardware devices so that you can organize your media collection and enjoy it. You can also stream TV shows, movies, music, pictures, and other streams on a television screen. However, while Kodi continues to grow in popularity, there are questions arising as to the legality of it and some of the add-ons. Some of these add-ons allow you access to content from YouTube, Netflix, and Hulu, to name but a few of the main media streaming services. The fact that Kodi is open source means that the door has been opened to some vulnerabilities and to questionable uses.

Chapter 6: Legal Issues

The Legal and Not So Legal Ways Kodi is Used

The obvious use is as a movie and TV streaming service and there is absolutely nothing wrong with this – provided you have a legitimate subscription to the service you are using. Even Amazon used to provide an app to let users install Kodi onto a fire stick or a Fire TV but that was pulled after Amazon decided that Kodi could be used for piracy and illegal content downloads. And there is no question of that because that is exactly what Kodi can be used for. Because it is open-source, anyone with a bit of knowledge can come up with add-ons and plugins that run a fine line between legal and illegal.

Many of the popular add-ons that allow you to watch paid streams without subscriptions are not really legal so, if you watch them without a VPN, you are taking the risk of being caught.

The Argument That Kodi is as Legal as Your Computer

While third-party add-ons certainly make things more complicated, they don't actually make Kodi illegal. The argument is that Kodi is as legal as your computer but it can also be used for illegal purposes. The actual technology does not violate any law anywhere in the world. As for the computer argument, there are two problems with this. Obviously, if you use a computer to do something illegal, it doesn't actually make the computer illegal but you can still get into trouble for doing what you do. It's the same with Kodi and this is what concerns many users.

Second, and a threat to the legality of Kodi, is that some versions of it have software snippets in without the use of a commercial license and these may be able to unscramble the system that a lot of DVD's are encrypted with – the Content Scramble System. This puts Kodi firmly in the field of violating the Digital Millennium Copyright Act, or DMCA.

The Argument That It Can't Be Illegal If Nothing is Downloaded

This is common but, sadly, some of the more powerful content providers, such as Amazon, Netflix, ESPN, HBO, etc., are not of the same view. If you don't pay for a subscription to their services, then the law is firmly on their sides. Earlier this years, the UK Premier League went after the Kodi users who stream their content without gaining permission. And there is the difference – streaming is not downloading.

Can I Get into Trouble if I Use Kodi?

The short answer to this would be yes. However, there are so many Kodi users now that it is highly unlikely that the authorities are going to take the time to trace everyone. Really and truthfully, the developers and the users who watch content that is copyrighted without the right permissions are violating laws and by streaming content, you are still violating those laws. Install a good VPN and you can use Kodi without any trouble.

In support of the legality of Kodi, it doesn't actually host content and there is no profit made from it. It operates under the General Public Licenses, which guaranteed freedom to copy, use and modify the Kodi software and it is also protected under the Communications Decency Act. Team Kodi will also remove add-ons that allow illegal streaming.

Conclusion

Thank you again for joining me in this book!

I hope this book was able to help you to understand how exciting Kodi is and what you can and can't do with it! There is no doubt that Kodi is the most popular media streamer in the world and, provided you set it up right, there really isn't any limit to what you can watch on it.

The next step is to scour the internet and look for other Kodi add-ons that can make your experience even better. There are plenty of forums and lots of help on the internet to ensure that all your questions are asked and that any problems you run into are easily solved.

Finally, if you enjoyed this book, then I'd like to ask you for a favor, would you be kind enough to leave a review for this book on Amazon? It'd be greatly appreciated!

Thank you and good luck!

www.ingramcontent.com/pod-product-compliance
Lightning Source LLC
Chambersburg PA
CBHW061201180526
45170CB00002B/901